A Force to Reckon With

Contents

Chapter 1

What Is All the Shaking About?

The day seems normal. The sky is blue and sunny. Suddenly, the ground starts to shake. Things start falling off the walls.

This is an earthquake. What causes the ground to quake?

A building damaged from an earthquake.

Earth has many layers. The top layer is called the **crust**. It is the thinnest layer. The crust is made of **plates**. These are sheets of rock. There are seven major and many minor plates in Earth's crust.

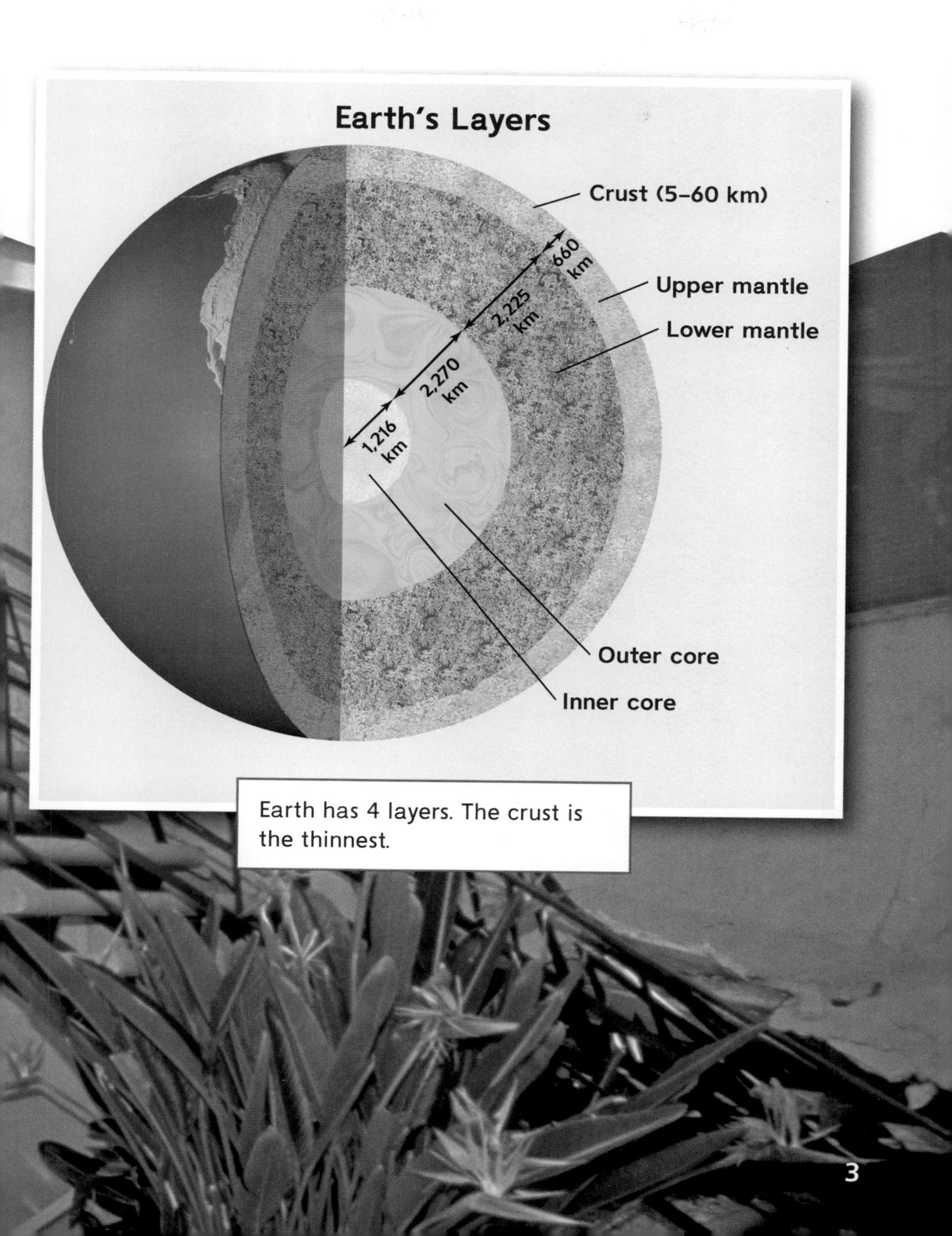

Earth has 4 layers. The crust is the thinnest.

The plates of Earth's crust are like puzzle pieces. All of the plates (pieces) fit together to make up Earth's crust (puzzle).

A puzzle's pieces fit together. But there are little cracks where the pieces come together. On Earth's crust, these cracks are called **faults**.

A fault line in Alaska

Earth's plates are always moving. The movement is slow. Earth's plates move at a rate of 0.66 to 8.50 centimeters each year.

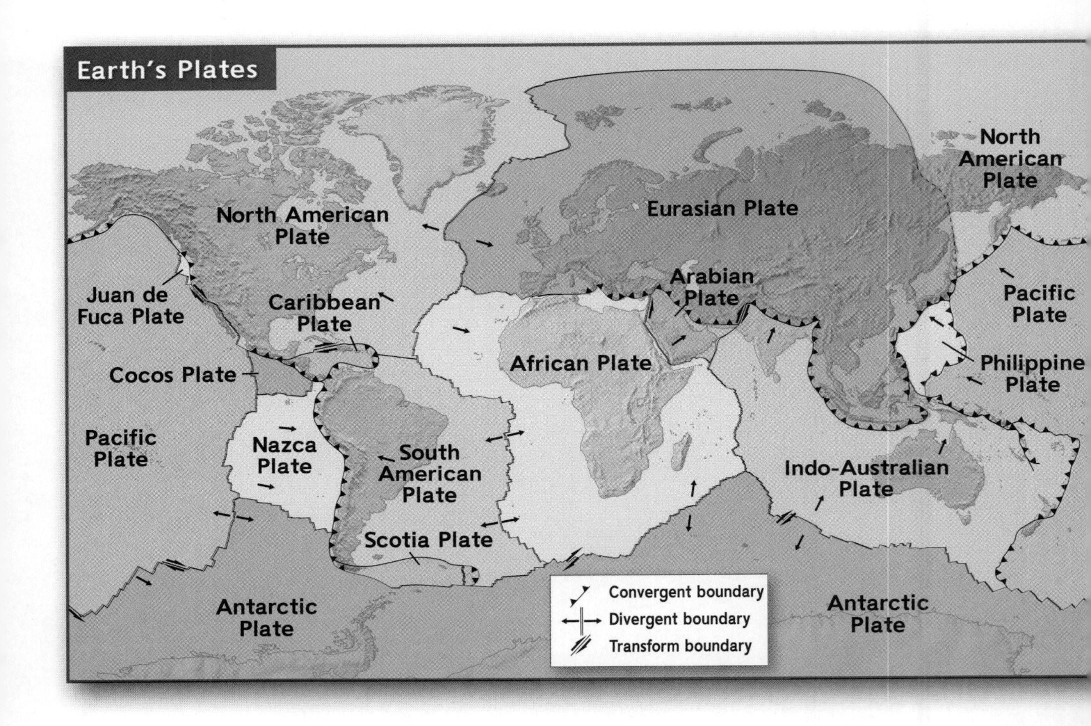

People cannot predict an earthquake based on the weather. There is no pattern between earthquakes and sunny, cloudy, or rainy days.

Earthquakes may occur when plates bump into each other, move apart, or slide past each other. Most earthquakes occur at the edges of the plates. This is where the plates meet.

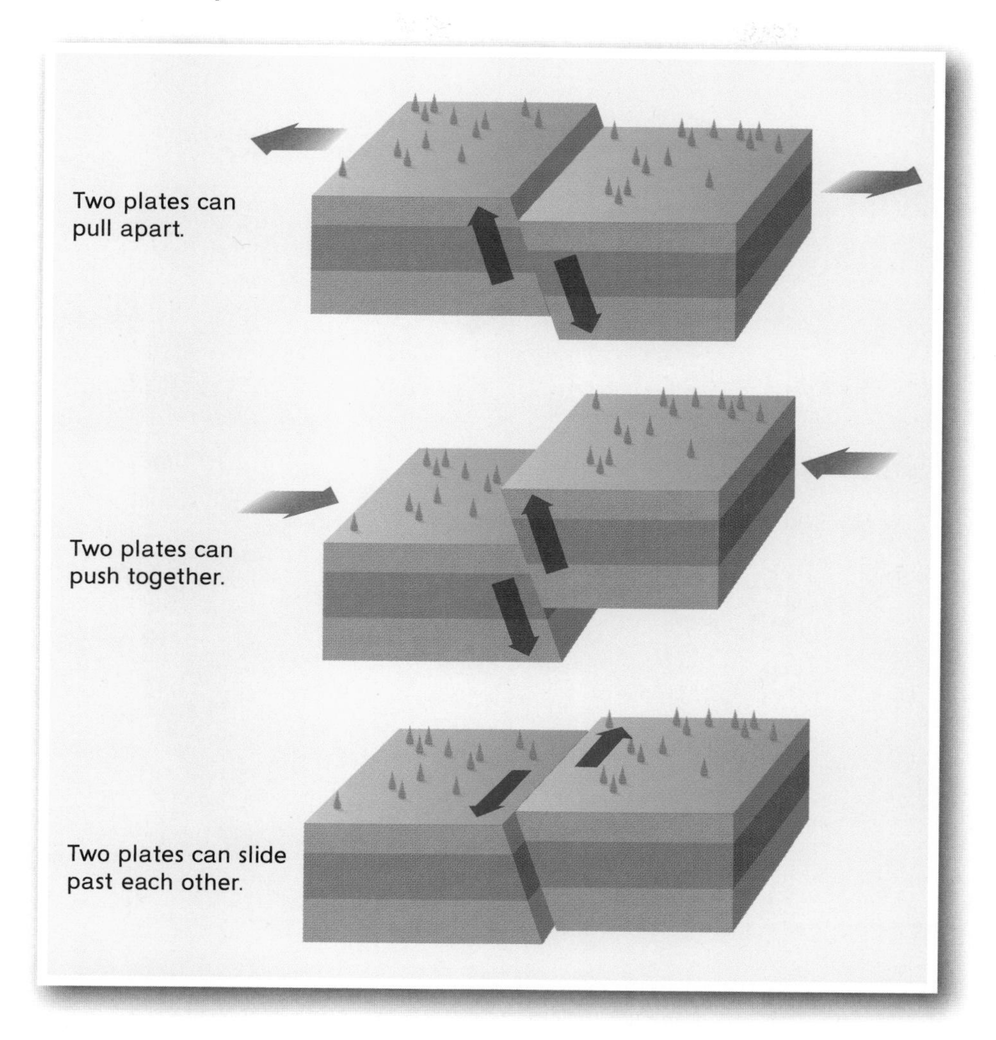

The place underground where plates meet is an earthquake's **focus**. An earthquake also has an **epicenter**. The epicenter is the point on Earth's surface directly above the focus. An earthquake's **radius** is the distance between the focus and the epicenter.

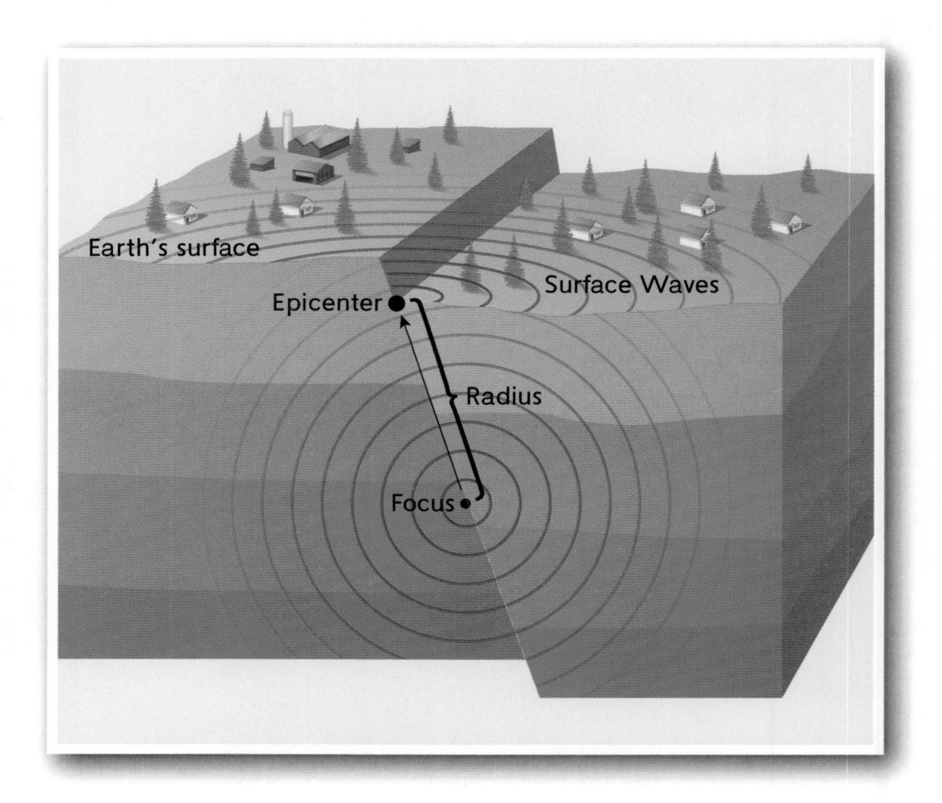

Chapter 2

Quake Belts

An area with a lot of earthquake activity is called a quake belt. The area covered by the Pacific Ocean plate is the largest quake belt on Earth. 80% of all earthquakes happen where the Pacific Ocean plate meets the plates that are under the continents along the Pacific Ocean.

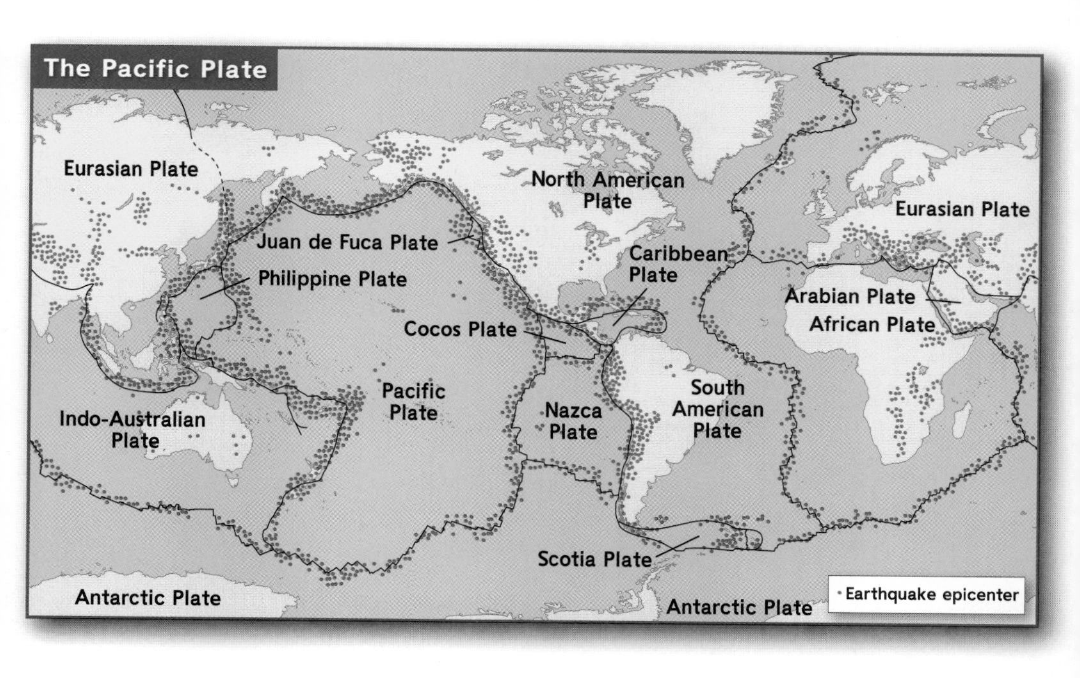

States affected by the Pacific Ocean quake belt are Alaska, California, Hawaii, Oregon, and Washington.

The top ten most powerful earthquakes that have happened in the United States were in Alaska. Some of the most powerful earthquakes in the United States were also in California.

Ten Most Powerful American Earthquakes			
Ranking	Location	Year	Magnitude/ Richter Scale Rating
1	Prince William Sound, Alaska	1964	9.2
2	Rat Islands, Alaska	1965	8.7
3	Andreanof Islands, Alaska	1957	8.6
4	East of Shumagin Islands, Alaska	1938	8.2
5	East of Shumagin Islands, Alaska	1946	8.1
6	Denali Fault, Alaska	2002	7.9
7	Gulf of Alaska	1987	7.9
8	Andreanof Islands, Alaska	1986	7.9
9	Rat Islands, Alaska	2003	7.8
10	Andreanof Islands, Alaska	1996	7.8

To learn more about the Richter scale, see page 16.

Florida and North Dakota have had the least number of earthquakes in the United States.

The San Andreas Fault is the main fault of the Pacific Ocean quake belt. It can be seen from Earth's surface.

The San Andreas Fault is more than 600 miles long. It stretches along the California coast. It is at least 10 miles deep.

The San Andreas Fault is located in CA.

The San Andreas Fault caused one of the most powerful earthquakes ever. It was the San Francisco earthquake of 1906.

In the 1906 earthquake, 300 miles of ground cracked along the San Andreas Fault.

The San Francisco earthquake was one of the worst natural disasters in history. The earthquake happened at 5:12 A.M. on April 18, 1906. It scored a 7.8 on the Richter Scale.

Many buildings were destroyed in the 1906 quake.

Did You Know?

Fire probably caused more damage to the city of San Francisco in 1906 than the earthquake itself.

Chapter 3

The Richter Scale

The Richter (RIK tur) scale was invented by Charles Richter in 1935. The scale measures the power of earthquakes. Earthquakes are rated on a scale of 1 to 10. The most powerful earthquakes have the highest ratings.

Richter Scale Magnitude			
Level	**Magnitude/ Richter Scale Rating**	**Description**	**Occurrence**
Micro	Less than 2.0	Undetectable by humans	Approximately 8,000 each day
Very Minor	2.0–2.9	Usually not noticeable, but recorded	Approximately 1,000 each day
Minor	3.0–3.9	Usually felt, but not strong enough to move	Approximately 49,000 each year
Light	4.0–4.9	Noticeable movement, and objects may fall from shelves and tables	Approximately 62,000 each year
Moderate	5.0–5.9	Buildings at the epicenter may experience damage	800 each year
Strong	6.0–6.9	50 miles from the epicenter can be dangerous, especially in cities	120 each year
Major	7.0–7.9	Expect destruction 100 miles from the epicenter	18 each year
Great	8.0–8.9	Buildings can be damaged a couple of hundred miles from epicenter	1 each year
Rarely, Great	9.0 or greater	Overwhelming damage can occur thousands of miles from the epicenter	1 every 20 years

The table describes Richter scale ratings and how often each type of earthquake occurs.

The Richter scale uses decimal points in its ratings. This allows scientists to provide more exact ratings.

Charles Richter

Math DETECTIVE

Look at page 16. What type of earthquake falls between a 5.0 and 5.9 rating? Which type of earthquake occurs most frequently? How do you know?

The Richter scale measures the power of earthquakes. A machine called a seismograph (SIZE muh graf) measures the vibrations made by an earthquake. The power of these vibrations is rated on the Richter scale.

This seismograph recorded an earthquake in California that had a rating of 6.5 on the Richter scale.

Chapter 4

The Biggest Quakes

The 1906 San Francisco earthquake caused a lot of damage. However, more powerful earthquakes have happened around the world.

Ten Most Powerful Earthquakes in World since 1900			
Ranking	Location	Year	Magnitude/ Richter Scale Rating
1	Southern Chile	1960	9.5
2	Prince William Sound, Alaska	1964	9.2
3	Off the West Coast of Northern Sumatra, Indonesia	2004	9.1
4	Kamchatka, Russia	1952	9.0
5	Off the Coast of Ecuador	1906	8.8
6	Rat Islands, Alaska	1965	8.7
7	Northern Sumatra, Indonesia	2005	8.6
8	Andreanof Islands, Alaska	1957	8.6
9	Assam, Tibet	1950	8.6
10	Kuril Islands, Japan and Russia	1963	8.6

Chapter 5

Predict and Prepare

Scientists study earthquake patterns. They learn about the history of earthquakes in certain areas. Scientists also study how Earth's plates move. This information helps them to predict when and where earthquakes might happen.

A scientist studying a crack in Earth's crust. It was caused by an earthquake.

Things can be done to prepare for earthquakes. Many new buildings are now built to be safe during earthquakes. People can make their homes safe in many ways.

Ways to Make a Home Safe for an Earthquake

- Fasten shelves securely to walls.
- Place large and heavy objects on lower shelves.
- Place breakable items such as bottles, foods, glass, and china in low, closed cabinets using latches.
- Make sure overhead light fixtures are secure.
- Repair defective electrical wiring and leaky gas connections.
- Get expert advice if there are signs of structural defects.

Did You Know?

Many cases have been cited where animals and pets change behavior just before an earthquake occurs.

People who live in areas with a high risk for earthquakes should have safety drills and emergency plans.

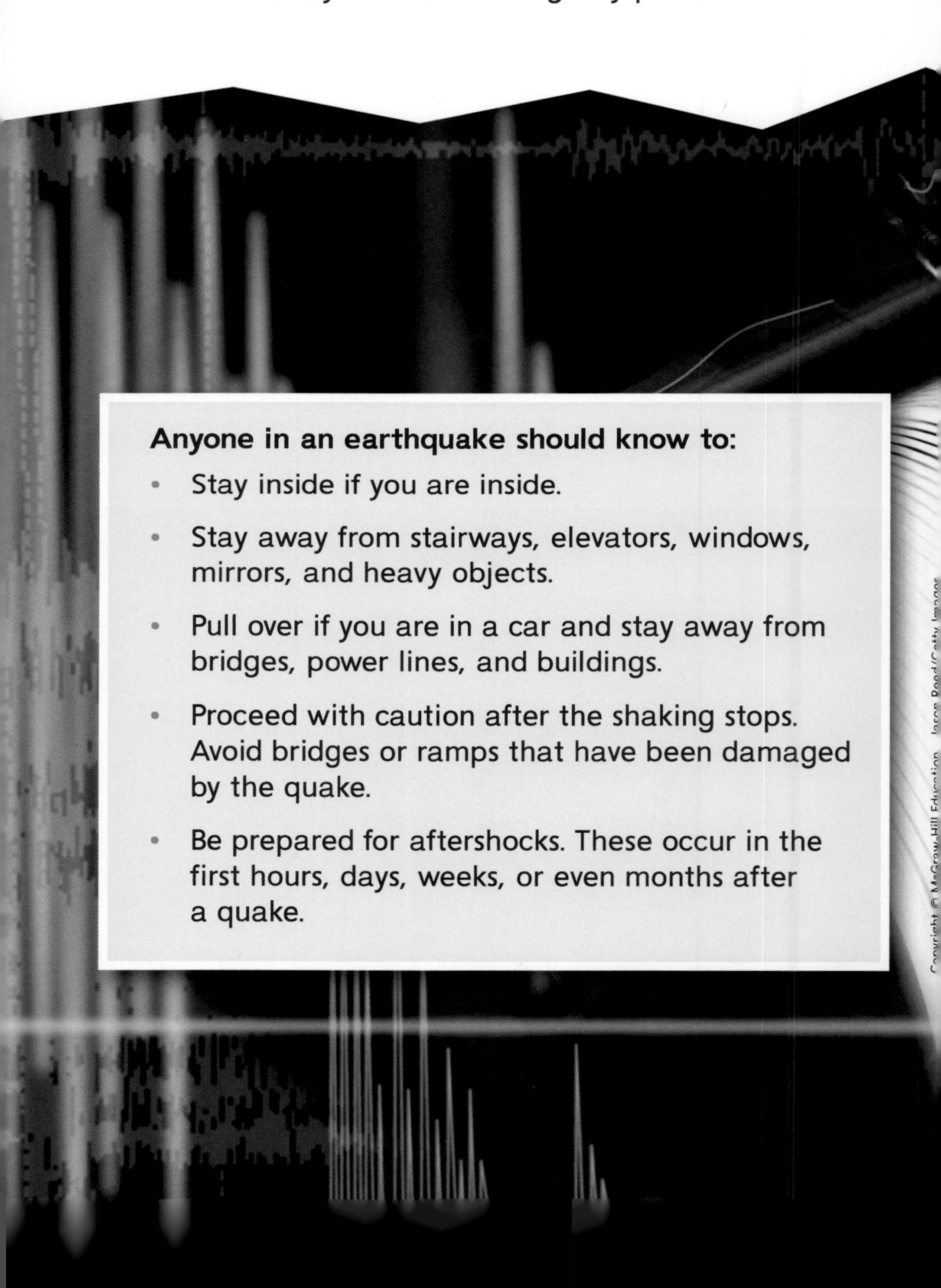

Anyone in an earthquake should know to:

- Stay inside if you are inside.
- Stay away from stairways, elevators, windows, mirrors, and heavy objects.
- Pull over if you are in a car and stay away from bridges, power lines, and buildings.
- Proceed with caution after the shaking stops. Avoid bridges or ramps that have been damaged by the quake.
- Be prepared for aftershocks. These occur in the first hours, days, weeks, or even months after a quake.

A person should have an emergency kit ready. The supplies needed in the kit are listed in the table below.

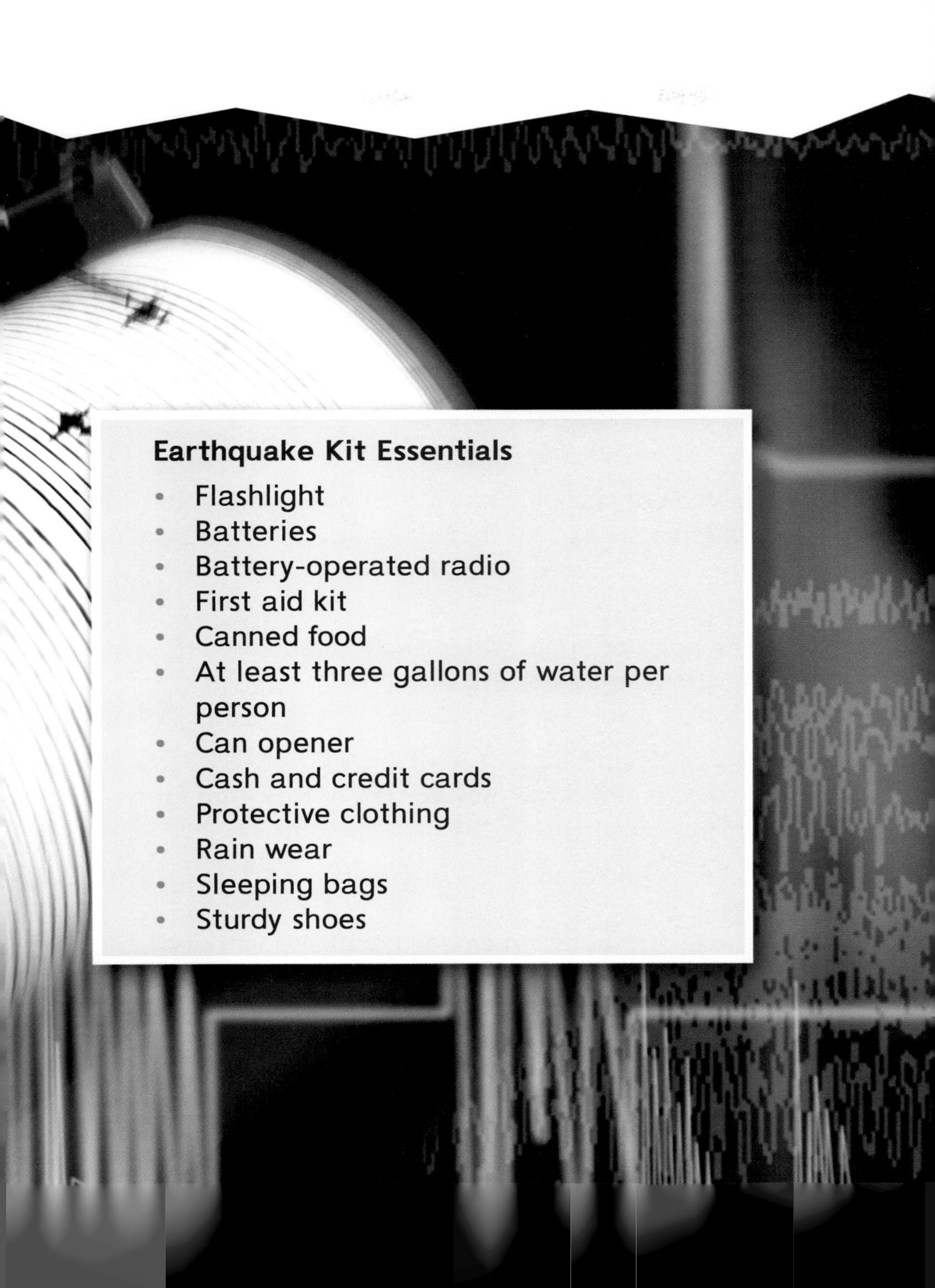

Earthquake Kit Essentials

- Flashlight
- Batteries
- Battery-operated radio
- First aid kit
- Canned food
- At least three gallons of water per person
- Can opener
- Cash and credit cards
- Protective clothing
- Rain wear
- Sleeping bags
- Sturdy shoes

Glossary

crust
The outermost layer of Earth. *(page 3)*

epicenter
The exact location on the Earth's surface directly above the focus of an earthquake. *(page 8)*

fault
A break or crack in Earth's crust where two plates come together. *(page 5)*

focus
The point of origin within Earth of an earthquake. *(page 8)*

plate
An extremely large moving slab of rock that forms Earth's crust. *(page 3)*

radius
The length between an earthquake's focus and epicenter. *(page 8)*

Real-World Problem Solving

1. Look at page 3. Earth's crust makes up $\frac{1}{4}$ of its layers. *One out of four* is represented by 0.25 in decimal form. What decimal represents the part of Earth's layers that is *not* crust? How else can you represent this relationship? [Chapter 1]

2. Look at page 6. Earth's plates move at a rate of 0.66 to 8.50 centimeters each year. What is the difference in the least and greatest distance Earth's plates can move in a year? [Chapter 1]

3. Look at page 11. What is the difference in magnitude of Alaska's first and tenth most powerful earthquakes? [Chapter 2]

4. Look at page 16. Which two categories have the most occurences? How does the magnitude of each category compare? [Chapter 3]

5. Look at page 19. What is the difference in magnitude of the United States' first and second largest earthquakes? Tell how you know. [Chapter 4]

GR P • Benchmark 38 • Lexile 750

Math and Science

Oceans: Into the Deep

Strange But True

Solving the Pyramid Puzzle

Ancient Giants of the Forest

Trapped in Tar

A Force to Reckon With

What Is Recycling?

Math and Social Studies

Rivers and Mountains of the United States

The Olympic Games

Americans on the Move

Class Project

Expanding the United States

Riding the Mail Trail

Life in the United States

Growing Goods in a Growing Country

ISBN: 978-0-02-100899-5
MHID: 0-02-100899-X

THE FOREST FARMER'S HANDBOOK

A Guide To Natural Selection Forest Management

Orville Camp